THIS CANDLEWICK BOOK BELONGS TO:

_____

_____

_____

_____

*For*_____

my brother Barrie

First U.S. paperback edition 1996

The Library of Congress has cataloged
the hardcover edition as follows:

Kitchen, Bert.
When hunger calls / Bert Kitchen.—1st U.S. ed.
ISBN 1-56402-316-8 (hardcover)
1. Predatory animals—Juvenile literature.
2. Predation (Biology)—Juvenile literature.
 [1. Predatory animals.] I. Title.
QL758.K57   1994
591.5'3—dc20   93-32360

ISBN 1-56402-971-9 (paperback)

10 9 8 7 6 5 4 3 2 1

Printed in Hong Kong

This book was typeset in Tiepolo-Book.
The pictures were done in acrylic paints.

Candlewick Press
2067 Massachusetts Avenue
Cambridge, Massachusetts 02140

WHEN HUNGER CALLS

BY BERT KITCHEN

CANDLEWICK PRESS
CAMBRIDGE, MASSACHUSETTS

When hunger calls, the killer whale may leave the sea . . .

The killer whale is found in all the oceans of the world, mostly around coastal waters but also out at sea. Its huge body and bold black-and-white markings make it easy to recognize. It is the largest member of the dolphin family.

Killer whales often hunt in packs, preying on seals, sea lions, porpoises, large fish, and other whales and dolphins. Sometimes a killer whale will even pursue its prey onto dry land—especially when sea lions are congregating on the South American coast to breed. The whale swims rapidly along sandy channels toward the beach and leaves the water to snatch its victim. Then, by twisting its body and making rapid movements against the sand, it is able to return to the sea with its prey.

When hunger calls, the great black-backed gull can hunt on the wing . . .

This large seabird can be found in many coastal regions, especially along the rugged coastline of the Atlantic Ocean. It lives largely on a diet of fish and vegetation but will also kill weak birds and mammals. It has been known, for instance, to snatch puffin chicks from the mouths of their burrows and is also agile and fast enough, in spite of its large body and wingspan, to capture adult puffins in flight. When a powerful gull dives down and snaps at the neck of the puffin with its strong beak, its small, short-winged prey stands little chance of escape.

When hunger calls, the angler fish casts its bait . . .

There are many different kinds of angler fish. This species, also known as the monkfish, can live at depths of more than three thousand feet in waters from the Barents Sea to North Africa. The angler fish moves slowly along the seabed, well camouflaged by its surroundings. It has a long, fleshy-tipped spine on top of its head and uses this as a lure to catch small fish, squid, and crustaceans. It waves the spine gently above its mouth to entice its victim close enough to reach, then snaps it up, trapping it in its hundreds of teeth.

When hunger calls, the osprey can catch its prey in one fell swoop . . .

The osprey is found in most parts of the world. Although it is in fact a member of the falcon family, it is sometimes called the fish hawk, preying on pike, trout, carp, bream, roach, and salmon. The osprey hovers and circles high above likely fishing waters. When it spots a fish, it sweeps back its wings and dives downward. At the last moment it brings its head up, thrusts its talons forward, and slows itself down by spreading out its wings. Sometimes the osprey goes fully underwater to reach its prey. Then it rises up again, flapping its wings to shake off the spray, with the fish hooked tightly in its talons.

When hunger calls,
a vulture has
no need to kill . . .

The vulture shown here belongs to an Old World
species that inhabits the plains of east Africa.
A vulture hardly ever kills for food, preferring to eat animals that are
dead. Its eyesight is excellent, enabling it to scan the landscape for signs
of carrion from high in the sky. Once it has discovered food, others will
often join it: A large carcass may be attended by fifty birds or more.
Vultures are able to digest even long-dead carrion. Although
their eating habits may seem repulsive to us, they actually
perform a very useful service in clearing away animal
remains from the territories they inhabit.

When hunger calls, the African rock python can swallow a gazelle . . .

The rock python lives in the southern and tropical regions of Africa and northward as far as the Sahara. It can grow to be over twenty feet long and is one of the biggest snakes in the world.

When the python is ready to strike, its head darts forward, and it seizes its prey in its jaws, then coils its body around it, either squeezing it to death or suffocating it. Like almost all snakes, the rock python can unhitch the two halves of its lower jaw and detach these from the two ends of its top jaw. This ability, together with the fact that the tissues of its throat are highly elastic, allows it to swallow animals much wider than itself. It can take over half an hour for the python to swallow an animal as large as this gazelle—and several weeks to digest it.

When hunger calls, the mongoose is a match for the cobra . . .

The mongoose of India and Southeast Asia
preys on insects, scorpions, small mammals, and snakes.
It is renowned for its snake-fighting ability.
A mongoose and a cobra, if they meet, will try to prey on
each other—and the mongoose usually wins. The cobra tries
to attack the mongoose with its poisonous fangs. These fangs
are quite short, however, and cannot easily penetrate the
mongoose's thick fur. Again and again the cobra strikes, but the
mongoose confuses it by twisting and turning like a boxer.
Once the cobra is exhausted and has wasted most of its venom, the
mongoose will lunge at its head or neck and deliver a fatal bite.

When hunger calls, the South American horned frog opens wide . . .

All frogs are carnivores, but most live on insects. Only the larger species, like this one, will eat lizards, birds, and small mammals as well. The South American horned frog's hunting technique is simple but deadly. It lies perfectly still, partially hidden by loose ground vegetation and foliage, and waits for prey. When a likely victim appears, the frog grabs it, crushes it with its strong jaws and sharp, stabbing teeth, and swallows it whole. A rat is probably the largest creature the horned frog will eat—though this species has been known to bite the lip of a grazing horse.

When hunger calls, the octopus makes use of its many arms . . .

The common Atlantic octopus is found in waters from the Mediterranean to the English Channel and from the West Indies to Connecticut. It lives on the seabed and makes its home in a reef or crevice, or underneath a rock. In spite of its frightening appearance, it is actually a shy and retiring creature.
The octopus crawls on its eight arms but can also propel itself along by forcing water through its body. It feeds on crabs, lobsters, fish, molluscs, and crustaceans, stalking prey until it is within easy reach, then wrapping its arms around it and smothering it. The octopus then takes the prey in its beaklike mouth (which is under its head, at the point where all its arms meet), poisons it with saliva, and crushes it. More often than not, the octopus takes a meal back to its den to eat at leisure.

When hunger calls, the spitting spider fires out glue . . .

The spitting spider is found all over the world.
It is tiny—less than a quarter of an inch long—and has
a highly effective way of catching food.
When it gets close enough to its prey—often a fly—the spider
raises its head and shoots two streams of sticky gum from the
tiny holes above its fangs, shaking its head from side to side so
that the gum forms a zigzag pattern and glues the victim to the
ground. The spider then closes in on it and gives it a fatal,
poisonous bite. Once the fly is dead, the spider removes
the strands of sticky silk from it before starting to eat.

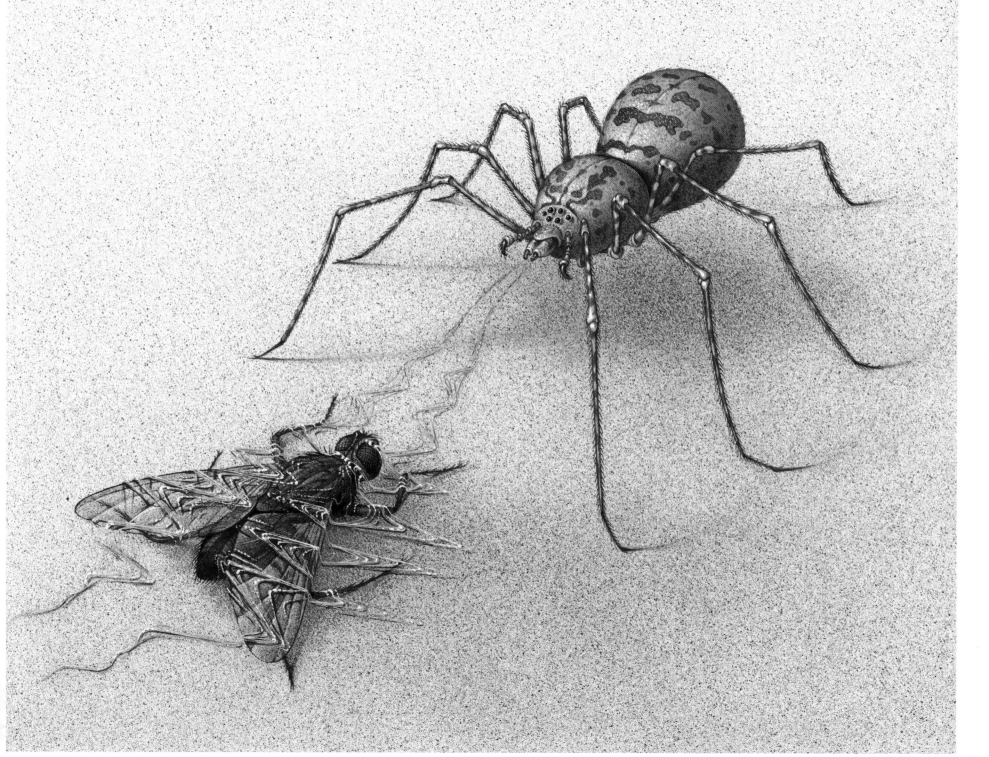

When hunger calls, the long-eared bat is listening . . .

The long-eared bat is widespread throughout Europe and Asia. A nocturnal mammal, it roosts in trees and buildings and—in winter—caves. Like most bats, it feeds on insects, spiders, and night moths, which it usually catches in midflight. To find its way around in the dark and to locate potential prey, a bat emits ultrasonic squeaks and clicks. The sound waves it produces bounce off any nearby object and are picked up again by the sensitive ears of the bat, allowing it to judge the exact size, distance, and direction of everything around it. Once it has tracked down a moth, the bat homes in on it, bringing its tail membrane forward to form a pouch and pocketing its prey to eat later.

And the woodpecker finch will even use a tool, when hunger calls.

The woodpecker finch belongs to a small group of birds known as Darwin's finches, found in the Galàpagos Islands. Although it eats some fruit, it lives mainly on insects and grubs. It does not have the long beak or tongue of a mainland woodpecker, but it does have its own unique way of getting food out of inaccessible places. If it discovers a grub buried in bark, for instance, it finds a cactus spine or small twig and uses it to extract its prey.

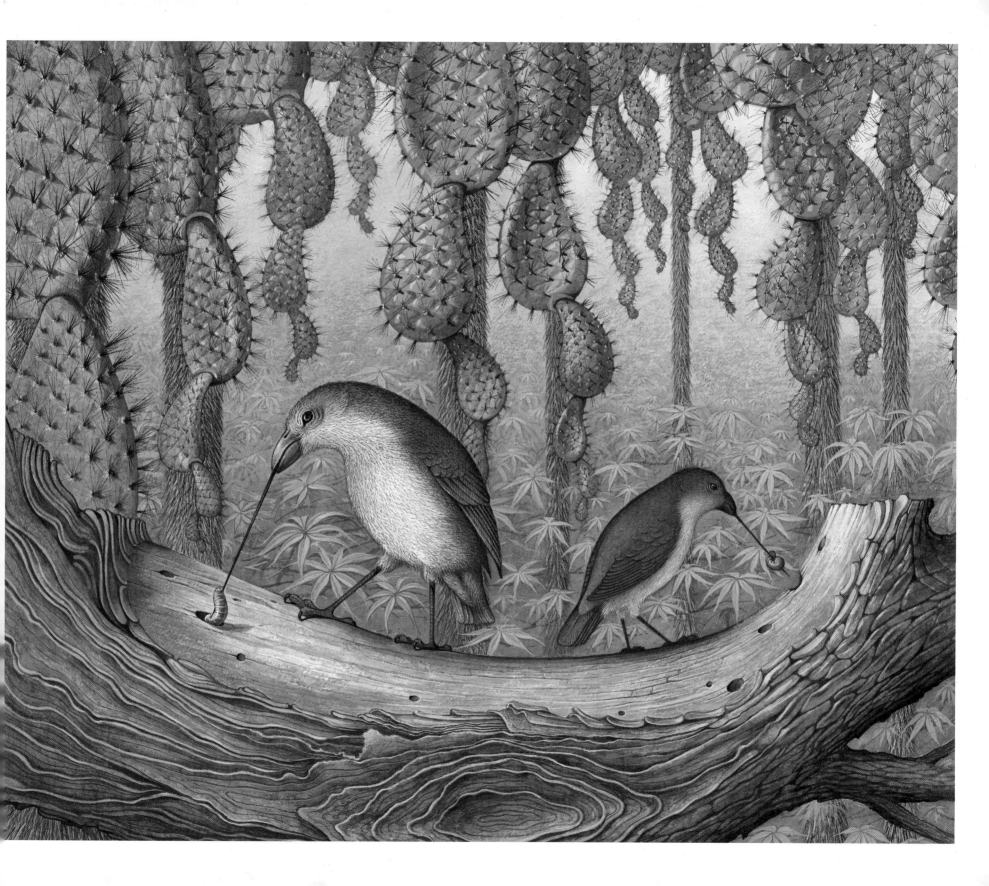

BERT KITCHEN, having completed his trilogy of picture books about animal behavior that began with *Somewhere Today* and *And So They Build . . .,* reports that this book was perhaps the most intriguing to work on of all. "There are so many fascinating and strange methods of predation. And food, of course, is as essential as a mate or shelter."